Understanding for

Graham Hill
Nick England

Series editor: Graham Hill

Contents

1 Push, pull, stretch and squash 4
2 Force and weight 6
3 Measuring and balancing forces 8
4 Friction – good or bad? 10
5 Slowing down 12
6 Speeding up 14
7 Under pressure 16
8 Pressure all around us 18
9 Twisting and turning 20

Forces are pushes and pulls. Forces get things moving or stop them, or make them change direction

Activities

1 Types of force 22
2 What do forces do? 22
3 Stopping distances 23
4 Stretching springs 24
5 Water supply 24
6 Designing a parachute 25
7 Turning forces 26
8 Glass bottles or plastic bottles? 26
9 What makes a bridge strong? 27
10 What are bridges made of? 28
11 How strong is hair? 28
12 Bridging the gap 29
13 Why do steel ships float on water? 29
14 Examining wind speed 30
15 Living in space 31
16 Forces – word quiz 32
17 Newton's ideas about motion 32

Acknowledgements

We are grateful to the following companies, institutions and individuals who have given permission to reproduce photographs in this book:

Action Plus Photographic (6, right; 8, top left; 11, middle right; 14, bottom left; 22, top middle, middle right; 25); Allsport/Vandystadt (4, bottom left); Heather Angel (17, middle left); Ardea London/Y Arthus Bertrand (14, middle; 22, middle bottom); Barnaby's Picture Library (6, bottom left; 13, middle left); Dr. Alan Beaumont (1, far right; 22, middle left); J Allan Cash (1, bottom left; 18, top; 19, left; 30; 31); European Space Agency (8, bottom); Leslie Garland (27, left and right; 28, left; 29); Sally and Richard Greenhill (22, bottom right); Hoverspeed (11, middle bottom); London Fire and Civil Defence Authority (19, top); Mansell Collection (6, top left; 28, middle); Metropolitan Police (23); NHPA (10, middle); Roddy Paine (4, bottom; 5, right; 8, middle, right; 10, top; 13, top right, middle right; 22, bottom left; 24; 26); Photographer's Library (4, top left); Pilkington (18, bottom); Popperphoto (7); Science Photo Library (5, left); Swan/John Buckingham (17, bottom); Swan/Grant Demar (16, left); Colin Taylor Productions (1, middle right, middle left; 4, top right; 10, bottom; 11, far left, top right and left, bottom right; 12, top and bottom; 13, far left, bottom; 14, bottom, top left; 15; 16, right; 17, top right; 20, top and bottom; 22, top left, top right; 28, right).

British Library Cataloguing in Publication Data
England, Nick
Understanding forces.
1. Forces
I. Title II. Hill, Graham, III. Series
531.11

ISBN 0 340 52519 3

First published 1990

Typeset and illustrated by Gecko Ltd, Bicester, Oxon
Printed in Great Britain for the educational publishing division of Hodder and Stoughton Ltd, Mill Road, Dunton Green, Sevenoaks, Kent by Cambus Litho, East Kilbride

Introduction

Science Scene is a series of books that will help your studies in Key Stage 3 of the National Curriculum.

Each book in the series looks at one of the attainment targets in science. This book – *Understanding forces* – covers the knowledge and understanding in Attainment Target 10 (AT 10 – Forces). As you work through this book you will be able to put the ideas that you read about into practice by doing activities. These will help you to develop some of the skills that are useful in your science studies. The activities will also help you to see how this attainment target overlaps with some of the others that you will meet in Key Stage 3.

There are two parts to the book:

- The first part contains short sections of reading with questions and Things to do. These sections will help you to understand the important ideas in this topic by reading, thinking and doing. You will know about some of these situations and contexts already. You will read about how forces play a part in all sorts of situations and contexts. They may occur at school, at home or around where you live. Other situations may be new to you.
- In the second part of the book there are lots of activities. These will give you another chance to learn about forces, but this time by doing. In other words, you will be investigating, thinking and finding out for yourself. Some of the activities are often quite long, and they can be difficult. So, your teacher will help you to choose the activities which are best for you.

If you work through both parts of the book, you will have covered almost all the work that is needed for AT 10 at Key Stage 3. In addition to this your teacher will plan practical work in the laboratory.

We hope that you will enjoy working through *Understanding forces*, and that it will make your science studies exciting and interesting.

Graham Hill and Nick England (1990)

1 Push, pull, stretch and squash

The team push hard to give themselves a good start on the bobsleigh run

Starting and stopping

You probably know already what a push or a pull is. When you are playing around at school you push people away from you, or pull them towards you. If you are standing still, a push or pull may start you moving. If you are already moving, a push or a pull may make you move faster or slow you down. When you are running, a push or a pull on your side makes you swerve or change direction.

Children pushing and pulling in the playground

Pushes and pulls are forces

You put forces onto lots of different things everyday. You use forces to move around. You pull yourself up a rope with your arms. You use forces when you squeeze the toothpaste out of a tube or when you turn on a tap. If you are good at a sport, or any activity, you will have learnt already how to use forces precisely. This might mean plucking a guitar, pushing a hockey ball, cutting out a paper model or lobbing a tennis ball.

The goalkeeper is about to make a save. The ball may not slow down but it will be pushed away from the goal. The push will make the ball change direction

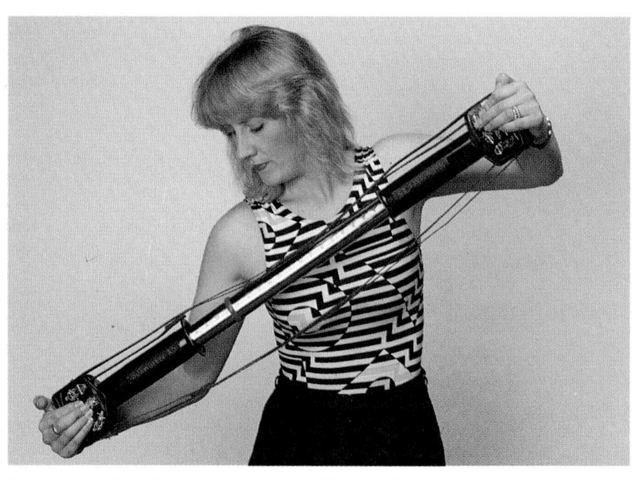

Anne applies force to stretch the spring

Stretching and breaking

Forces can do other things besides making objects start and stop moving. Forces make things squash, stretch and break. Stretch a piece of Plasticine or a rubber band. These materials change their shape when a force acts on them. When the force is removed, materials like Plasticine stay out of shape. They are called **plastic materials**. Other materials like rubber return to their original shape when the stretching force is removed. These are called **elastic materials**. Have you ever tried to stretch a piece of china or glass? If you did, you would find that these materials do not stretch. China and glass break under large forces, but they don't change shape under small forces. We call them **brittle materials**.

You can use force to change the shape of clay. Is clay elastic or plastic?

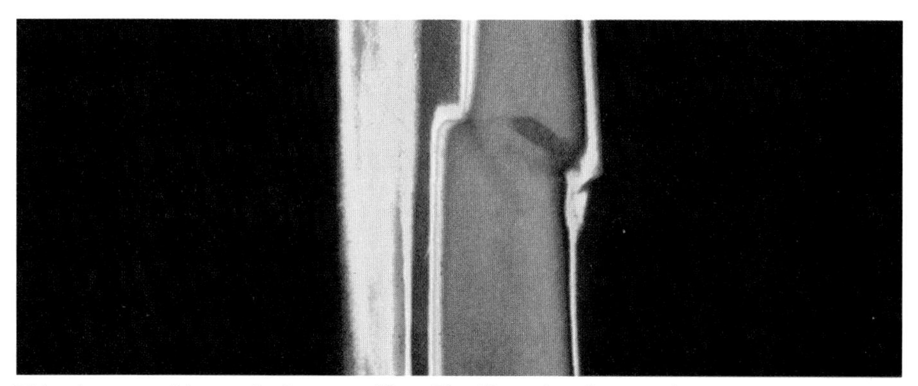

This shows an X-ray of a fractured leg. The X-ray has been coloured so that the broken bone shows up more clearly

Remember that:

- a push or a pull is a force
- pushes and pulls can start or stop you moving
- pushes and pulls can speed you up, slow you down or make you swerve
- you need forces to stretch, squash or break things.

Things to do

1 Look at the bobsleigh in the photograph. What forces are used to get the bobsleigh moving at the start of the run? What forces make it swerve as it goes down the run? What forces make it stop?

2 Get into a group with two or three friends. Think of things that you do everyday which need forces. Make a list of examples where you use a pulling force, and another list of examples of using a pushing force. What effect do these forces have? Do they speed something up or stretch it or do something else?

3 a) Name a plastic material besides Plasticine.
b) Name an elastic material besides rubber.
c) What is the difference between plastic and elastic materials?

4 Plan an experiment to investigate how the length of an elastic band depends on the force applied to it.

2 Force and weight

Isaac Newton was born in 1642. He made important discoveries about motion, forces, gravity, light and colour

There is one force that acts on you and everything else all the time. When you drop something, the object falls to the ground. When you trip up, you land on the floor. The force which pulls our bodies towards the ground is called the force of gravity. This force is pulling you right now.

The pull of **gravity** on you is called your weight.

Isaac Newton was the first scientist to realise that there are forces of attraction between all objects in the universe. He called these forces of attraction **gravitational forces**. The gravitational force between two people, or between you and a house is too small to notice. But the pull between you and something as large as the Earth is very strong.

The force of gravity helps this scuba diver to get down to the seabed

The polevaulter is pulled down to the ground by the force of gravity

Animal	Weight
humming bird	0.02 N
mouse	1 N
cat	30 N
adult woman	600 N
adult man	700 N
bull African elephant	50 000 N
blue whale	2 000 000 N

Table 1

What is your weight?

Forces are measured in newtons. This name was chosen in honour of Isaac Newton. Weight is a force, so it is measured in newtons. The table shows the weights of some animals. Your weight is probably about 500 newtons; this is usually written as 500 N. The symbol N stands for Newton. As you can see from table 1, the larger the weight, the bigger the pull of gravity.

Ed Aldrin enjoying low gravity on the moon, 21st July 1969. Without his specially weighted suit, Ed would find it difficult to control his movements

Losing weight

The photograph above shows Ed Aldrin walking on the moon. On Earth, dressed in his space suit he weighs about 1800 N. On the moon he weighs only 300 N. He weighs less on the moon because the moon pulls him less than the Earth. The moon is smaller than the Earth so it pulls with a weaker gravitational force.

As a spacecraft travels away from the Earth, the pull of gravity from the Earth gets less and less. Once it is a long way from the Earth, the pull of gravity on it gets too small to notice. On the way to the moon Ed Aldrin and his companions felt weightless, because they could not feel any gravitational pull from either the Earth or from the moon.

Things to do

1 Suppose your friend has been away from school. How would you explain to him or her what we mean by the force of gravity?
2 Ed Aldrin lost weight when he went to the moon. Explain why. Had he been on a diet?
3 'What goes up must come down.' Discuss this statement in a group with two or three others. Is this saying always true?
4 Write an account of a day in your life as a 'weightless' astronaut. What problems will you have?
5 Ceres is an asteroid, or minor planet, which moves round the sun in an orbit. Its diameter is 900 km. This is about ¼ of the moon's diameter. Estimate what you think Ed Aldrin would weigh on Ceres. Explain how you made your estimate.
6 Use books and encyclopaedias to find out about the life and work of Isaac Newton. Concentrate on the work that Newton did on motion, forces and gravity. Write a short article about Newton, using your findings.

3 Measuring and balancing forces

This gymnast weighs 400 N What force does she need to pull herself up on the bars?

The kitchen scales work by using a spring balance

This angler is using a spring balance to weigh his fish

Gases leaving the rocket's exhaust push the rocket upwards with a force of 10 million newtons.

In the last unit you learnt that forces are measured in newtons. Your weight is a force, and its size is about 500 N. How big are other forces? How hard can you push or pull?

You can push or pull with a force of a few hundred newtons. When you push someone in the playground, you probably use a force of about 100 N. Your foot kicks a ball with a force of about 300 N, and your arms can pull on a rope with enough force to lift you up. Machines can push or pull with much larger forces. A tractor can pull with a force of 10 000 N. It takes a force of 100 000 N to lift a small aeroplane off the ground.

Forces and weights can be measured using spring balances. These are marked in newtons. Because of this, spring balances are sometimes called newton-meters — look at the photos above. Newton-meters work because forces on the spring will either make them longer or shorter.

Balanced forces

You know that forces can change the motion or the shape of something (page 5). When the motion is changed, an object speeds up, slows down or changes direction. But sometimes when you apply a force to something, there is no change in motion or shape. When a force causes no change, it must be balanced by another force. Forces that act in opposite directions at the same point and that balance each other are called **balanced forces**.

Figure 1 shows some examples of balanced forces. When Mark sits at his desk (Figure 1a), the force of gravity pulls him downwards. But, he does not move because another force balances his weight. The chair is pushing up on him. Because he does not move, the upwards push from the chair must be as strong as the downward pull of gravity. The upward push from the chair and the pull of gravity are balanced forces. Figure 1a shows how the forces on Mark can be drawn as arrows. The arrows are the same length because the forces are the same size, but the arrows point in opposite directions.

Figure 1b shows two Sumo wrestlers, Taiho and Toshimitsu, locked in combat. At the moment, neither is succeeding in pushing the other back. The forces on each wrestler balance. Taiho's weight is balanced by an upward push from the floor. Taiho also feels a sideways push from Toshimitsu. But, since he is not falling backwards, Toshimitsu's push is balanced by a push from the floor in the opposite direction.

Figure 1a

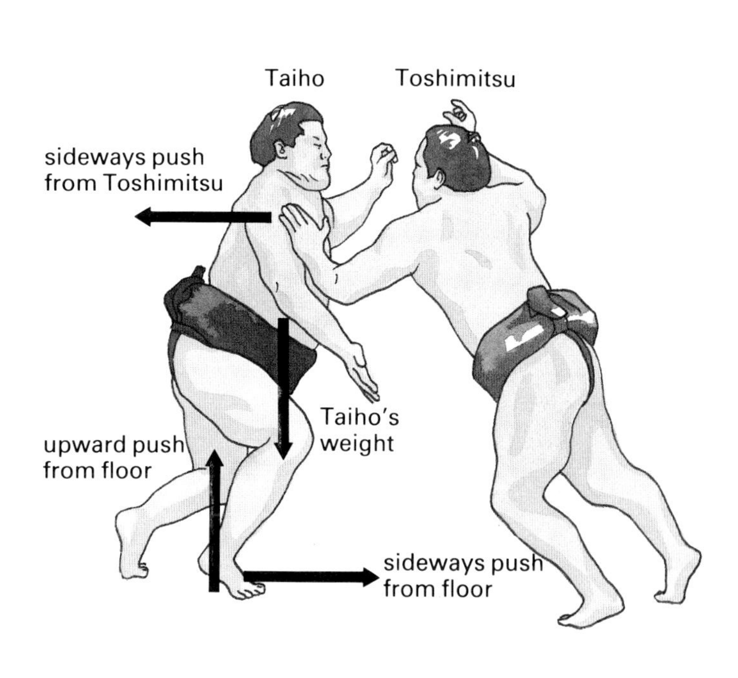

Figure 1b

Things to do

1 Make your own weighing machine using springs, rubber bands or luggage straps. Describe how you would use it to measure the weights of some objects. How would you use it to work out the weight of these objects in newtons?

2 Explain what we mean by *balanced forces*.

3 Forces can be represented by arrows. What do the length and direction of the arrows represent?

4 A parked car has balanced forces acting on it. Draw a diagram to show the balanced forces acting on the car.

5 Have a look at your kitchen scales at home. Press on them.

a) What do you think is inside the scales?

b) Try to explain how they work.

6 Look at figure 1b. Draw another diagram to show the balanced forces acting on Toshimitsu.

4 Friction – good or bad?

Have you ever tried to push a car along a flat road? If so, you will know how difficult it is. The car does not move easily or freely because of friction. Friction is a force. It is the force which acts when one surface moves over another. Friction slows down moving things like cars, bicycles and balls.

What are the effects of friction?

Friction slows down moving objects.

Look at the photo of the cyclist. If Annabel stops pedalling, she will slow down. This is because she is moving through air. This causes air resistance. There is also friction at the axles and on the chain. If Annabel did not keep on pedalling she would stop.

Friction can make things hot. These bushmen are using friction to start a fire

Friction causes wear and tear.

Have you noticed how quickly the soles of your shoes or trainers wear away when you are sliding, stopping and turning? Friction between your footwear and the ground causes this wear and tear.

Friction can make things hot.

Rub your hands together. After a short time, they get hot because the rubbing causes friction. In the same way, the moving parts of a machine can also get hot. The photo shows how bushmen use friction to start a fire.

Reducing friction

We can reduce friction in three ways.

- **Lubrication**
  We use oil to reduce friction between the moving parts of machinery. The oil is a lubricant. It works by separating the surfaces with a thin layer of liquid. Some machines use air as a lubricant. Air lubrication is used in hover mowers, hovercrafts and aerotrains. These move over land, over water or along rails on a cushion of air.
- **Ball Bearings**
  Look closely at the space between a bicycle front wheel and its axle. You should see a lot of small steel ball bearings (figure 1). These roll over as the wheel turns and this reduces the friction.
- **Streamlining**
  The air resistance on a fast moving car can be as much as 1000 N. The water resistance on a large speedboat is much more than this.

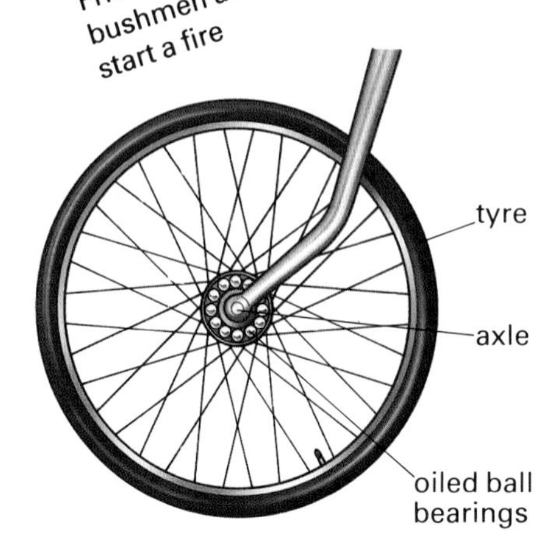

Figure 1 Reducing friction with oiled ball bearings

The resistance which fluids, like air and water, exert on moving objects is called *drag*. We can reduce drag by streamlining the shapes of cars, boats, trains and aircraft. Streamlined vehicles can travel faster and further on the same amount of fuel.

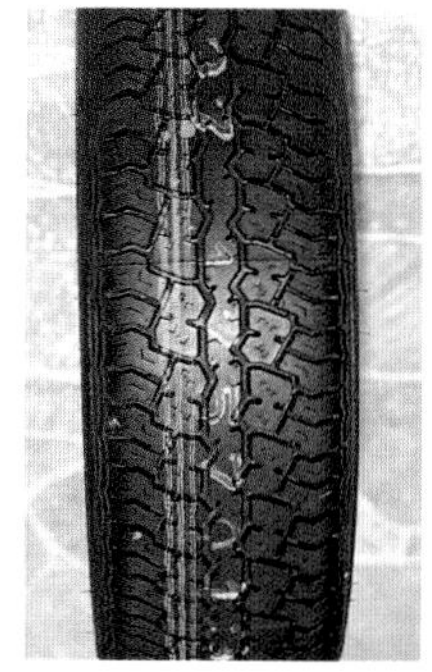

Look at the photos above. It is illegal to drive a car with less than 2 mm of tread on the tyres. When the tread on the tyres is thick, water on the road is forced into the tread. The tread can then grip the road. If the tread is thin, the water cannot get into the tread. In this case, the car slides on a thin layer of water

Examples of streamlining in technology

Although friction can be a nuisance, it is very necessary. Friction between our shoes and the ground allows us to stop and start and change direction. Walking on frictionless ground would be impossible. It would be worse than trying to walk on ice.

Friction is also important for vehicles to work. Without friction between the road and the tyres, it would be impossible to control a car. In wet weather, water on the road reduces friction and makes driving more dangerous. This is why drivers are warned to take more care.

The shape of a car is carefully designed to reduce drag

A skin-tight suit reduces the drag acting on the skater

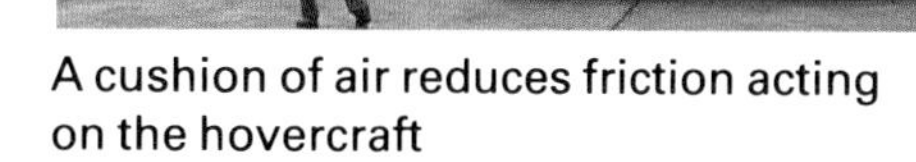

A cushion of air reduces friction acting on the hovercraft

Dolphins are mammals, like cats, dogs and humans. How are they streamlined for movement through the water?

Things to do

1 Explain what is meant by *friction* and *drag*.

2 a) What force pushes a yacht forward?

b) What forces of friction slow down a moving yacht?

3 a) Give three examples of movement in which friction is a problem.

b) How could the friction be reduced in each of these examples?

4 Prepare a talk to the rest of your class on **Streamlining in animals**. Your talk should last about 3 minutes. Collect photos or draw diagrams to illustrate your talk.

5 Slowing down

Useful information
50 mph = 80 km per hour
= 2.2 m per second

10 mph = 16 km per hour
= 44 cm per second

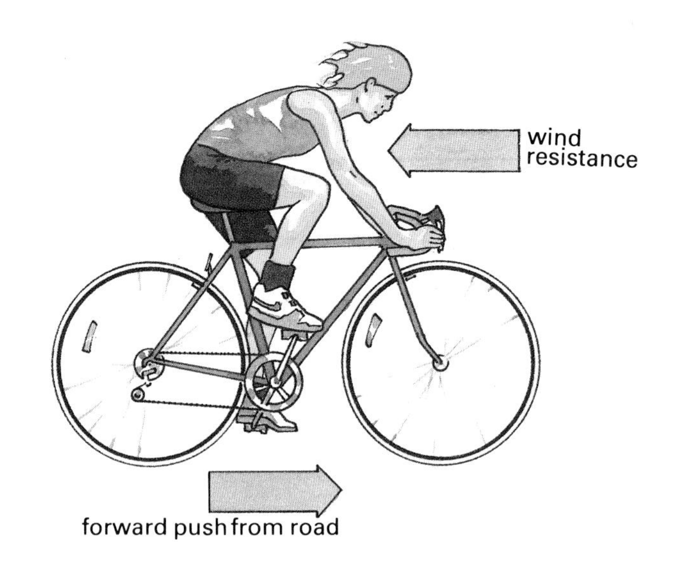

Figure 1 When a cyclist moves at a steady speed, the push from the road is balanced by wind resistance and other frictional forces

When you pedal your bike, you turn the back wheel. The back wheel pushes against the road, and the road pushes the bike forwards. The push from the road speeds you up. As you go faster you can feel the wind resistance. At a high speed, the push from the road is balanced by the wind resistance. This means that you cannot speed up any more. If you stop pedalling, the wind resistance will slow you down.

You can, of course, stop your bike by using the brakes. If you are moving slowly, you can stop your bike in a short distance. You only need to apply the brakes gently. But if you are moving fast, it takes longer to stop. You need to apply the brakes with more force and you also travel further before you stop. Look at the photos of the two cars on the left.

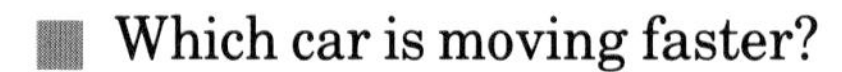

- Which car is moving faster?
- How did you decide which car was moving faster?
- Suppose both cars are equally heavy. Which car will stop first if both drivers put on their brakes with the same force?
- What must the drivers do to stop their cars even sooner?

Which of these cars is moving faster?

Approx speed in miles per hour	Speed of car in metres per second	Distance to stop after brakes are applied	Time to stop after brakes are applied
35	15	15 metres	2 seconds
70	30	60 metres	4 seconds

Table 1 The distances and times in which it takes a medium-sized car to stop after the brakes are applied

When a car is moving fast, it needs a large force to stop it. Fast-moving cars take longer to stop than slow-moving cars. Fast-moving cars also travel further in the time between applying the brakes and stopping.

Look at the information in table 1. This information is taken from the Highway Code. The first column in the table shows the speed of the car in miles per hour. This speed tells you how many miles the car will travel in one hour. The second column shows the same speed in metres per second. This tells you how many metres the car will travel in one second.

When the car is travelling at 35 miles per hour, it needs about 15 metres in which to stop once the brakes are applied. This distance is called the **braking distance**. The car will take about 2 seconds to travel this 15 metres. If the same car is travelling at 70 miles per hour, the braking distance is 60 metres. This time the car takes 4 seconds to stop.

These examples show that the faster you are moving, the more difficult it is to stop. It needs a greater force and a longer time. When you travel fast, you need a bigger distance in which to stop. This is why road signs warn drivers to keep to speed limits and to keep a sensible distance behind the vehicle in front.

This poster warns pedestrians not to step out in front of a car

This sign warns motorists that they are travelling too fast. It lights up when a car travelling too fast approaches

Warning stickers on the back of cars help to remind drivers to keep their distance

Drivers must keep their distance to prevent an accident like this one

Things to do

1 Imagine that you are going to buy a new bike. Make a list of the things that you think are important in choosing a bike. Compare your list with those of others in your class.

2 What two forces act on a car that is moving at a steady speed? Are these two forces balanced?

3 What happens to the braking distance of a car (i) when the road is icy, (ii) when the road is rough, (iii) when you are going downhill?

4 Why is it very important to slow down when it is foggy on a motorway?

5 Work in pairs. Plan an experiment to compare the brakes on different bikes. Say what you would do. Say what measurements you would make. Say how you would decide which bike had the best brakes.

6 Speeding up

Moving object	Speed (m/s)
glacier	0.000001
snail	0.0005
human walking	2
human sprinter	10
express train	60
Concorde	600
Earth moving round Sun	30 000
light and radio waves	300 000 000

Table 1

This snail will move 100 m in 2½ days

Lindford Christie won the Commonwealth Games 100 m in 9.93 seconds. What was his average speed?

Some things move very fast. Others move very slowly. Look at the wide range of speeds in table 1.

When you travel in a fast car you finish your journey quickly. When you travel in a slow car, your journey takes longer. Suppose that the speed of a fast car is 100 kilometres per hour. We write this as 100 km/h. The slanting line is short for 'per' or 'in one'. So a speed of 100 km/h means you travel 100 km in one hour, or 100 km per hour.

The simplest way to find your speed is to measure

- the distance you travel (or 'd' for short)
- the time you take (or 't' for short)

$$\text{Then, average speed} = \frac{\text{distance travelled}}{\text{time taken}} = \frac{d}{t}$$

We call this the average speed because the speed of the car may change during the journey. It will slow down when it is stuck behind a lorry and speed up to overtake a car.

Look at the cheetah in the photograph. She covers 100 m in 4 seconds. What speed is this in metres per second?

To catch her prey, this cheetah needs a rapid burst of acceleration

This sports car can reach a high speed in a short time. It accelerates quickly

Distance-time graphs

Sue lives in Crowthorne which is about 70 km from central London. When she leaves home to drive to London she has to drive 10 km to the motorway. Once Sue is on the motorway she can drive quickly. But the last 5 km into central London are very slow because there is a lot of traffic. Figure 1 shows a graph for Sue's journey. The graph shows how far Sue has travelled after different times. It is called a **distance-time graph**.

After 10 minutes Sue has travelled 10 km, after 40 minutes she has travelled 65 km.

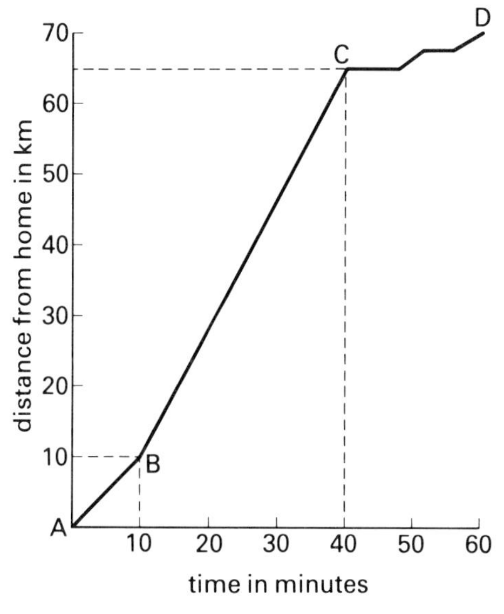

Figure 1 Sue's journey into London

Changing speed – acceleration

When a car is speeding up, we say it is **accelerating**. When it is slowing down we say it is **decelerating**. You can think of deceleration as a negative acceleration.

A car that accelerates rapidly reaches a high speed in a short time. For example, a sports car speeds up to 45 km/h in 3 seconds. A bus speeds up to 45 km/h in 15 seconds. The bus takes five times longer than the sports car to reach 45 km/h. So, its acceleration is five times slower. We can work out the acceleration of the car using the formula:

$$\text{Acceleration} = \frac{\text{increase in speed}}{\text{time taken}} = \frac{45 \text{ km/h}}{3 \text{ s}} = 15 \text{ km/h per second}$$

Every second, the car's speed increases by 15 km/h. Can you work out the acceleration of the bus?

This bus acclerates more slowly than a sports car

Things to do

1 Explain what the following terms mean. Use your own words.
(i) speed (ii) average speed (iii) acceleration
(iv) deceleration

2 This question is about Sue's journey to London. You will need to use the graph in Figure 1 to answer some of these questions.

a) How far was the journey?
b) How long did the journey take?
c) What was Sue's average speed for the whole journey?
d) The car was travelling fastest on the motorway when the slope of the graph is steepest. How far did Sue travel on the motorway?
e) What was Sue's speed on the motorway in km/h?
f) Look at part C–D of the graph. What do you think happened at this stage of the journey?

3 If a car travels at 60 km/h for 15 minutes, how far does it travel?

4 When Carl Lewis runs a 100 m race he accelerates rapidly from the starting blocks. He reaches a speed of 40 km/h after 2 seconds.

a) What is his acceleration?
b) Does the sports car described above accelerate more quickly than Carl Lewis?

7 Under pressure

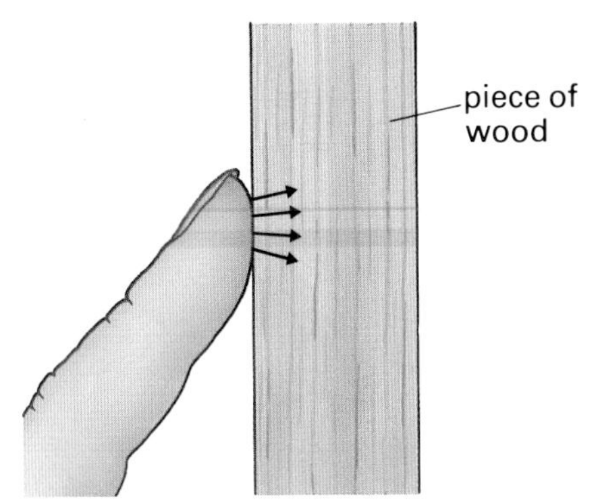

Figure 1a It is difficult to push your finger into the wood because the force is spread over a large area

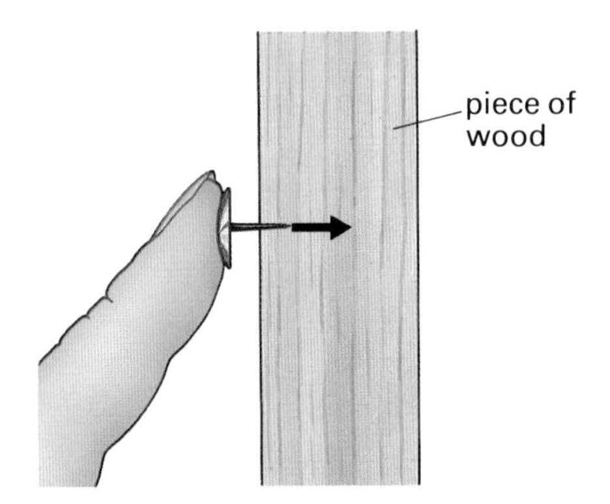

Figure 1b The force is concentrated in a small area so the drawing pin can be pushed into the wood

Look at figure 1. Why is it easy to push a drawing pin into wood, but impossible to push your finger, into the wood? When you press on the wood with your finger, the force is spread over a large area. But when you press onto the drawing pin, the force on the wood is concentrated on a small point. When the force is concentrated in this way, we say the pressure is high. Pressure tells you how concentrated a force is. It tells you the force on one square metre (m^2) or the force on one square centimetre (cm^2).

$$\text{Pressure} = \frac{\text{Force}}{\text{Area}}, \qquad \text{i.e. } P = \frac{F}{A}$$

The force is measured in newtons, and area is measured in square metres or square centimetres. So the pressure is measured in newtons per square metre (N/m^2) or newtons per square centimetre (N/cm^2).

Stiletto heels produce high pressure. They concentrate the weight of a person wearing them into a very small area. This makes it difficult to walk across grass without sinking. Stiletto heels also damage polished floors.

The 4 tonne elephant in the photograph weighs 65 times more than the woman wearing stiletto heels, but the pressure under his feet is less than that under hers. The soles of his feet are so large that his weight is spread over a large area.

The photograph across shows Nick lying on a bed of nails. How can he lie there without being hurt? Nails are sharp and can pierce the skin. However, Nick has spread his body out so that he lies on a lot of nails. The area of the nails supporting him is so large that it does not hurt (too much)!

If the elephant stood on your foot, would he do more damage to it than the woman in stiletto heels?

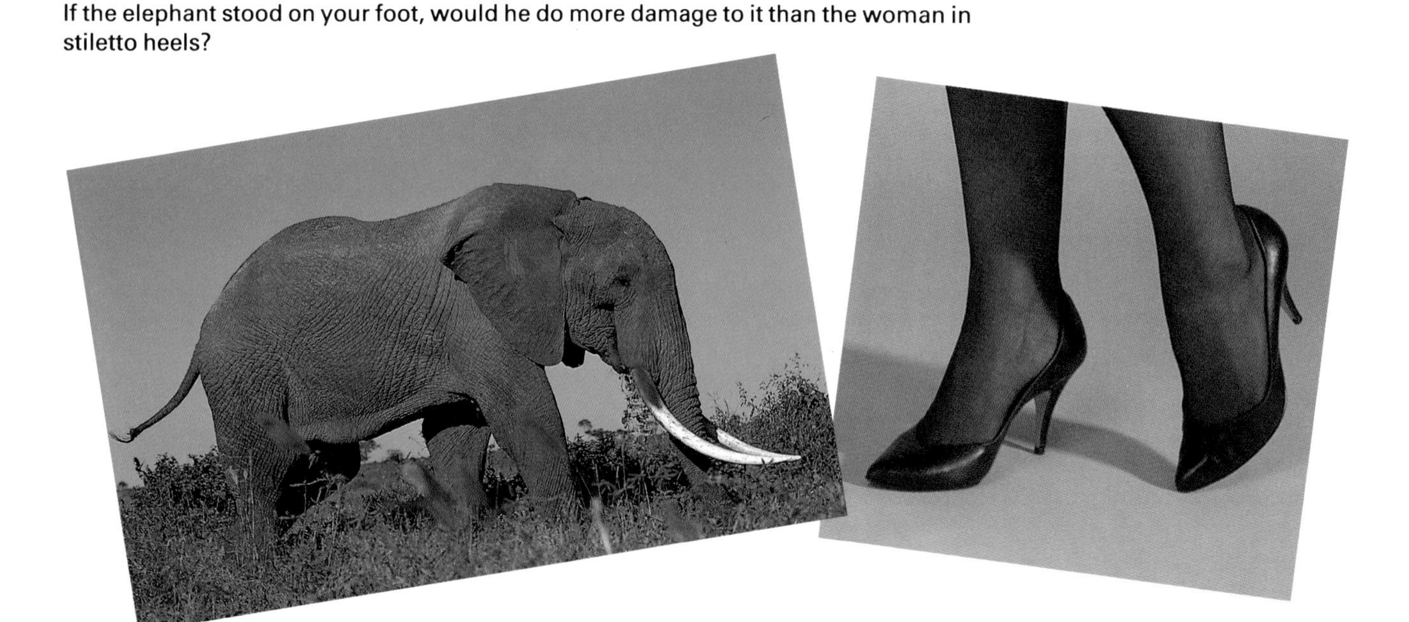

Nick lying on his bed of nails

Why do tractors need large, wide tyres on boggy ground?

Is it easier to walk over snow wearing skis or snow shoes?

Things to do

1 Inga, who weighs 500 N is a keen skier. Her shoes have an area of 0.02 m^2 in contact with the snow. Her skis have an area of 0.2 m^2 in contact with the snow.

a) What pressure does Inga put on the snow when she is wearing shoes?

b) What pressure does Inga put on the snow when she is wearing skis?

c) Why does Inga find it easier to walk over snow wearing skis rather than shoes?

2 Look at the photo of the woodpecker. Woodpeckers have very strong skulls and sharply pointed beaks. Why are these useful?

3 Work in small groups. Plan an experiment which you could do to find the weight of a car. (*Hint*: you will find it helpful to use the equation, pressure $= \frac{\text{force}}{\text{area}}$, and to measure the pressure in the car tyres.) If possible ask an adult if you can try out your experiment using his or her car.

Why does a woodpecker need a very strong skull and a sharply pointed beak?

8 Pressure all around us

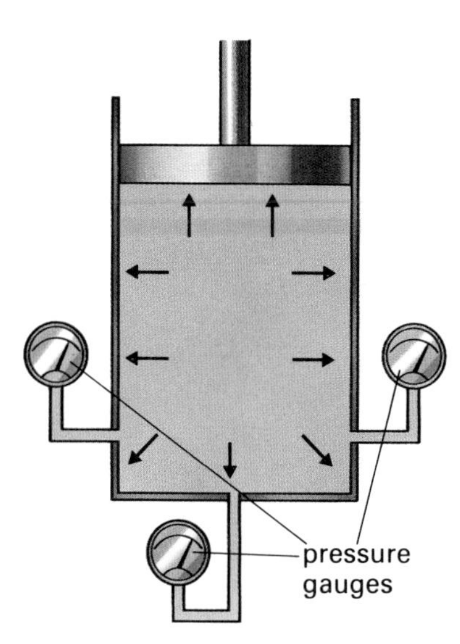

Figure 1 Pressures in liquids act in all directions — all these pressure gauges read the same

Pressure in water

When you hit a nail with a hammer, the pressure acts downwards through the point of the nail. This happens because the nail is a solid. Liquids behave differently. In figure 1, a cylinder of oil is squashed by the piston. The oil pushes outwards in all directions rather than in just one direction, like the point of the nail. So, the pressure is the same all around the cylinder.

Figure 2

When a deep-sea diver goes deeper and deeper, the pressure on him increases more and more. This is because there is more water above him. At a depth of 100 m in the sea, the pressure is 1 million newtons per square metre. Imagine 1000 large men piled on a square metre, as in figure 2. They would cause the same pressure. Deep-sea diving is very dangerous. Divers have to wear protective suits like the one in the photo, and they can dive only for short lengths of time. Next time you go to the swimming baths, try swimming along the bottom of the pool. You will feel the extra pressure as you dive down deeper. Dam walls are made thicker near the bottom because of the extra pressure in deeper water (see figure 2 on page 25).

This diver is wearing a special suit for working in deep water

Pressure in the air

The pressure from the air around and above us is also very large. Air pressure is sometimes called **atmospheric pressure**. Normally you do not notice air pressure because the pressure inside your body is the same as the pressure in the air around you. Atmospheric pressure is used to lift up heavy but fragile sheets of glass, as shown in the photograph. The method is explained in figure 3.

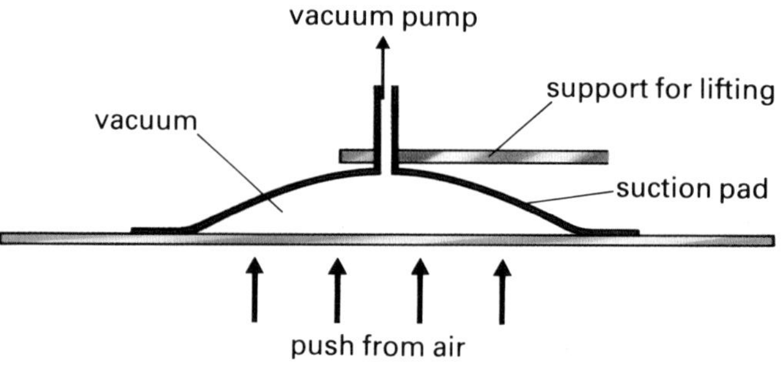

Figure 3 A vacuum pump sucks the air out of the suction pad. This makes a vacuum above the glass. A vacuum is a space with nothing in it. There is no air to push down on top of the glass. So, the air underneath the glass pushes the glass upwards onto the pad

We notice pressure changes in the air because they cause winds. If the pressure between two places changes gradually then there will be a gentle breeze as air moves from one place to another. For example, there is often a gentle sea breeze blowing from the sea to the land. If the pressure between two places changes quickly then there will be a strong wind.

This powerful water jet is caused by high pressure in the hose pipe

In January 1990, there were rapid changes in air pressure all over Britain. These created very strong winds and stormy weather. Thousands of trees were blown over and there was a lot of damage to buildings. Figure 4 shows how the air pressure in Oxford changed on January 25th 1990. The fastest winds in Oxford were at 3.15 in the afternoon when the air pressure was changing very quickly indeed. You can see this in figure 4. It is the steepest part of the graph.

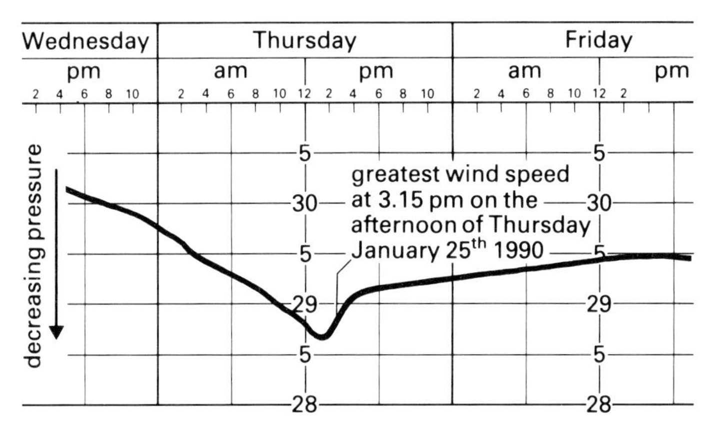

Figure 4 This graph shows how the atmospheric pressure dropped on a winter's day. The greatest wind speeds occur where the pressure changes most quickly

This photo shows some of the damage caused by the January storm. It takes a strong force to blow down a tree

Things to do

1 This is an experiment for outside. Get an old washing-up liquid bottle and fill it with water. Squeeze it gently, then squeeze it strongly. Why does the water come out more quickly when you press it hard? Now stick a skewer into the side of the bottle in two places as shown in figure 5. Fill it with water and then draw a diagram to show how the water flows out of the holes. Explain what you see.

2 Explain in terms of differences in pressure, how you can drink through a straw.

3 Watch the weather forecast on the television. Look at the lines on the weather map. (These are called **isobars**. They join up places with the same pressure.) What do you notice about the isobars when the wind is strong?

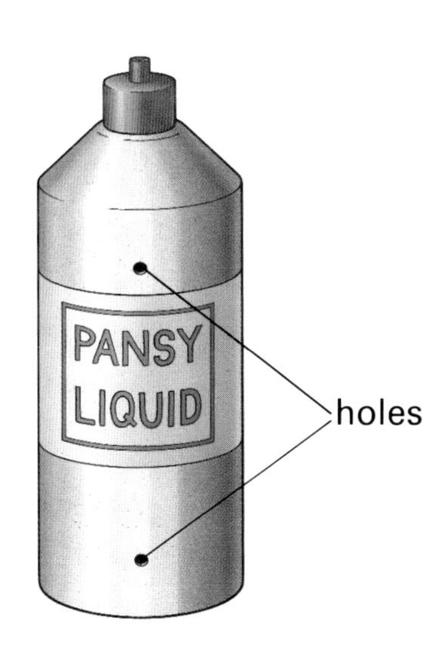

9 Twisting and turning

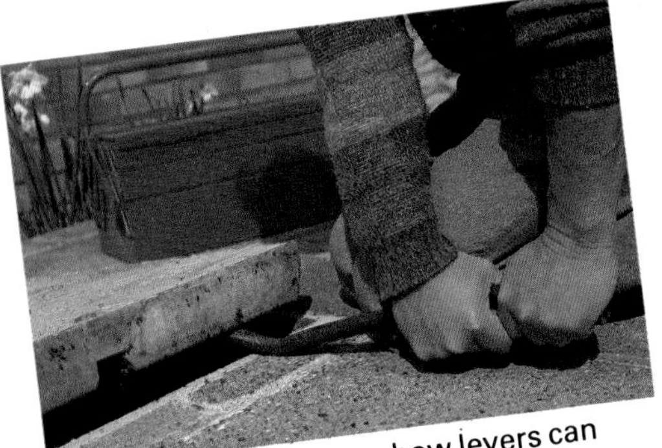

The photos above show how levers can be used to do different jobs

Hold a heavy book at arm's length. How long can you hold it there? Now hold the book close to your chest. How long could you hold it there? You could probably hold it there for hours. It is much easier to hold a book close to your chest than at arms length. The force of gravity pulling on the book causes a turning effect. The turning effect is much bigger when the book is held at arms length — see figure 1. We call this turning effect the **moment** of the force.

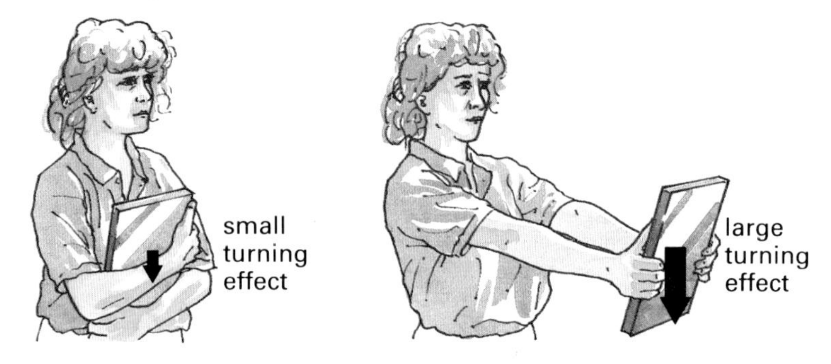

Figure 1 It is easier to hold a book close to your body than away from you

Tools for jobs

We use tools like spanners, levers and bottle openers to increase the turning effect of a force. They help us to do jobs more easily. It is impossible to undo a tight nut with your fingers. Using a spanner the job is easy, even though you use the same force. But you only get a large turning effect if you push in the right direction (figure 2). You get no turning effect if you push towards the nut.

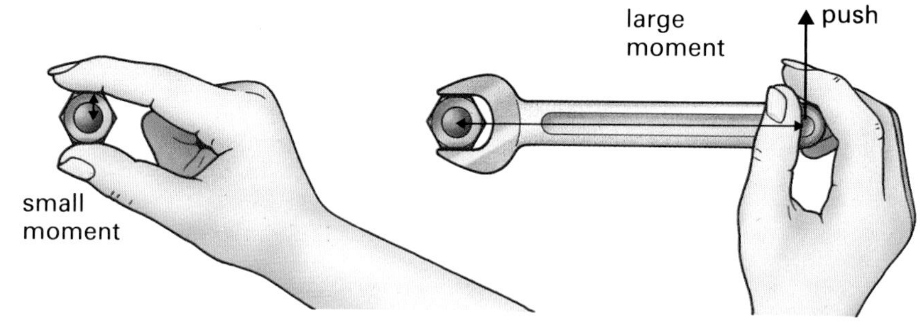

Figure 2

You can use screwdrivers and crowbars as levers. You can use a screwdriver to remove the lid from a tin of paint. The screwdriver turns (pivots) on the edge of the tin. The point where the lever turns is called the pivot or **fulcrum**.

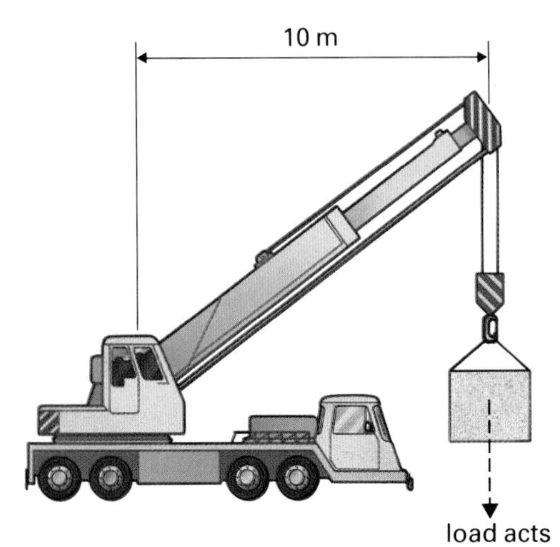

Figure 3

A mobile crane, like the one in figure 3, can help workmen to lift heavy loads. But if the turning effect of the load is too large the crane will tip over. If the distance from the cab to the load in figure 3 is 10 m, the crane can lift about 50 tonnes. But if the distance from the cab to the load is 20 m, the crane can lift only 25 tonnes safely. If the load is heavier than 25 tonnes, the crane will topple over. This is because the turning effect (moment) of the load is twice as large. Notice that the turning effect of a force depends on both the size of the force **and** the distance of its line of action from the fulcrum.

Moment of force = force × distance from fulcrum to line of action of force

Balancing

Look at figure 4. The see-saw acts like a lever. Vicky and Jaipal both produce a moment on the see-saw. In figure 4, Vicky's weight turns the see-saw down to the left. Jaipal's weight turns the see-saw the other way. The see-saw does not turn because the turning moments caused by their weights balance.

Moment of Vicky's weight = moment of Jaipal's weight

$$600\text{ N} \times 2\text{ m} = 400\text{ N} \times 3\text{ m}$$
$$1200\text{ Nm} = 1200\text{ Nm}$$

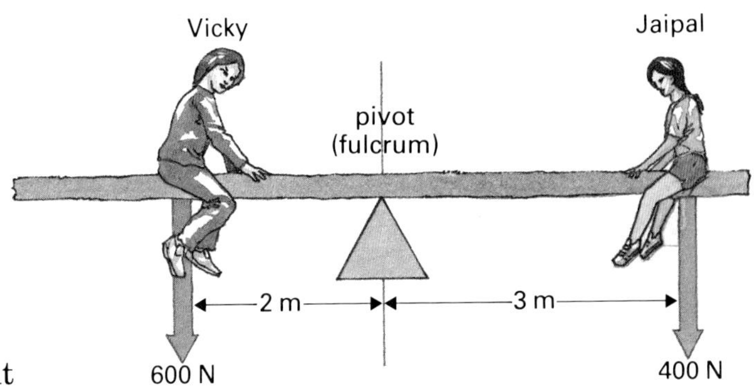

Figure 4

Things to do

1 a) When you cannot undo a tight screw, you use a screwdriver with a large handle. Explain why.
b) Explain why door handles are not put near the door hinge.
c) How could you increase the moment that you were using to tighten a nut?

2 Where do you have to put the spare penny in figure 5 to balance the ruler?

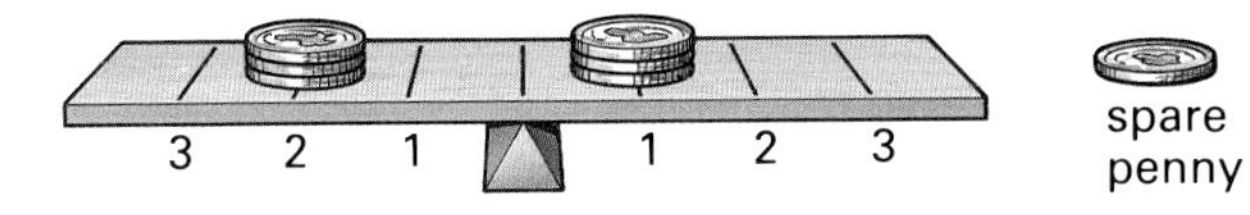

Figure 5

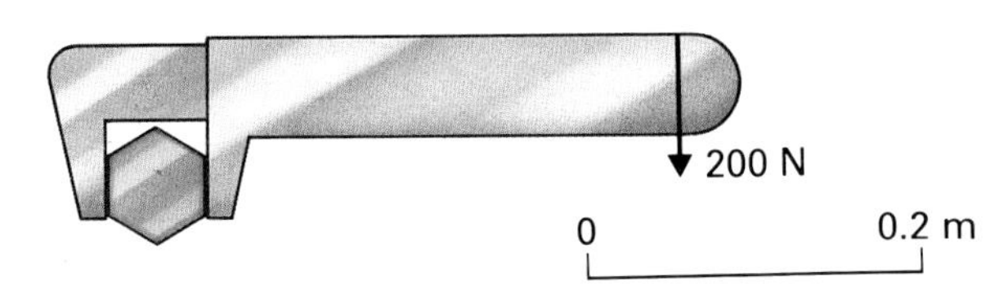

Figure 6

3 Look at figure 6. What turning moment about the nut, does the 200 N force have?

4 Make a pair of scales using a drinking straw balanced on a needle. Use your scales to compare the weights of small objects like pins and paper clips.

5 Look at the photograph of the steelyard below. This is a bit like a see-saw.
a) Where is the fulcrum?
b) What will tip the steelyard left?
c) What will tip the steelyard right?
d) Why is the right side of the beam so long?
e) How is it used to weigh things?

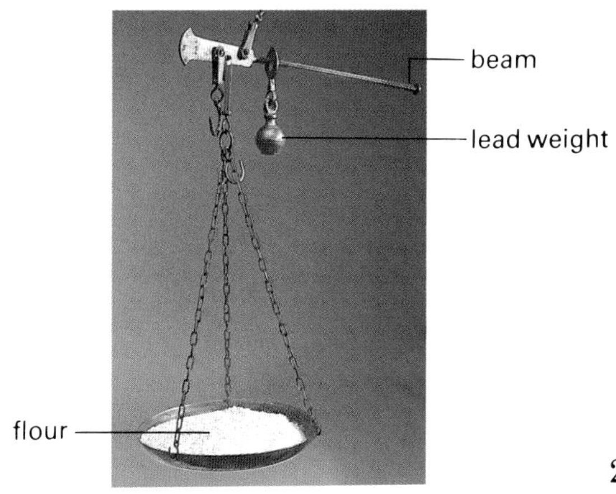

Activities

1 Types of force

There are different types of force. Pushes, pulls, twists and turns are four examples.

1 Look at these photographs. What kind of force is there in each photograph?

2 What is the effect of each force?

2 What do forces do?

If you wanted to put a force on something, you could push it, pull it, kick it, squeeze it, twist it, etc.

In this activity you will be putting forces on materials. This will help you to understand what a force is and what forces can do. Work in pairs and discuss the questions which follow the experiments.

Experiment 1

Try this experiment with three or four of the following materials: Plasticine, rubber or rubber band, sponge, match, polythene, paper clip, clay, cork, Blue-tac.

1 What happens when you put a force on the material? (You may have to try pulling it, pressing it, or bending it.)

2 What happens when you remove the force?
3 Do all the materials behave the same with roughly the same force?
4 Why is Plasticine described as *plastic*? How do plastic materials behave when forces are put on them?
5 Name one other plastic material.
6 Why is rubber described as *elastic*? How do elastic materials behave when forces are put on them?
7 Name one other elastic material.
8 What do forces do to the materials you have tested in this experiment?

Experiment 2

Put a marble on a flat surface.

9 What happens when you push it very gently?
10 Repeat the experiment, but this time push it with a stronger force. What happens?
11 Try the experiment again, but this time while the marble is moving, apply a force in the opposite direction. What happens?
12 Try the experiment again, but this time while the marble is moving, apply a force at right angles to the direction in which it is moving. What happens?
13 What do forces do to the marble in this experiment?

3 Stopping distances

Table 1 below is taken from the Highway Code. The information relates to a medium-sized car. (Some values have been left out.)

Speed in miles per hour	Thinking distance in feet	Braking distance in feet	Overall stopping distance in feet
20	20	20	40
30		45	75
40	40		120
50	50	125	175
60	60	180	
70	70	245	315

Table 1

1 The thinking distance is the distance travelled during the time that it takes for the driver to react. What is the braking distance?
2 What is:
(i) the thinking distance at 30 miles per hour?
(ii) the braking distance at 40 miles per hour?
(iii) the overall stopping distance at 60 miles per hour?
3 (i) How does the thinking distance change as the speed increases?
(ii) What happens to the thinking distance if the speed doubles?
(iii) How does the braking distance change as speed increases?
(iv) What happens to the braking distance if the speed doubles?
4 (i) Plot a graph showing the car's speed (vertical) against the overall stopping distance (horizontal).
(ii) What is the overall stopping distance at a speed of 55 miles per hour?
(iii) Suppose it is very foggy and you can see only 100 feet ahead. What should the maximum motorway speed be under these conditions?
5 Suppose a car is travelling on a wet, slippery road. How will this affect the thinking and braking distance?
6 Suppose the car driver is tired, but other conditions are good. How will this affect the thinking and braking distance?

A police car on a skid pan, where police learn to control a skidding car

4 Stretching springs

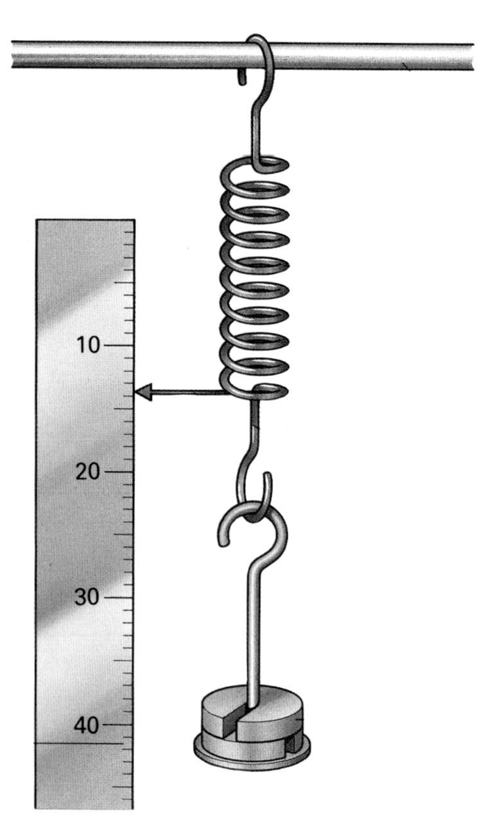

Figure 1

We can study the force of gravity by hanging weights from a spring. Gravity from the Earth pulls on the weights and this stretches the spring (figure 1).

A pointer is attached to the lower end of the spring. This is used to measure the extension of the spring when weights are attached to its lower end. Table 1 shows the scale readings of the pointer for different weights.

Weight attached in newtons	0	2	4	6	8
Scale reading in cm	13.0	13.7	14.4	15.1	15.8

Table 1 Studying the stretching of a spring

1 Why must the spring be attached to a rigid support?
2 Use the results in table 1 to make a table showing the stretching force on the spring, in newtons, and its extension (or stretch) in cm.
3 Plot a graph of the stretching force (vertical) against the stretch (horizontal).
4 How does the stretch of the spring depend on the stretching force?
5 You can use your graph of force against stretch to weigh various objects. What is the weight of a can of baked beans which stretches the spring by 1.7 cm?
6 How does a spring balance work?

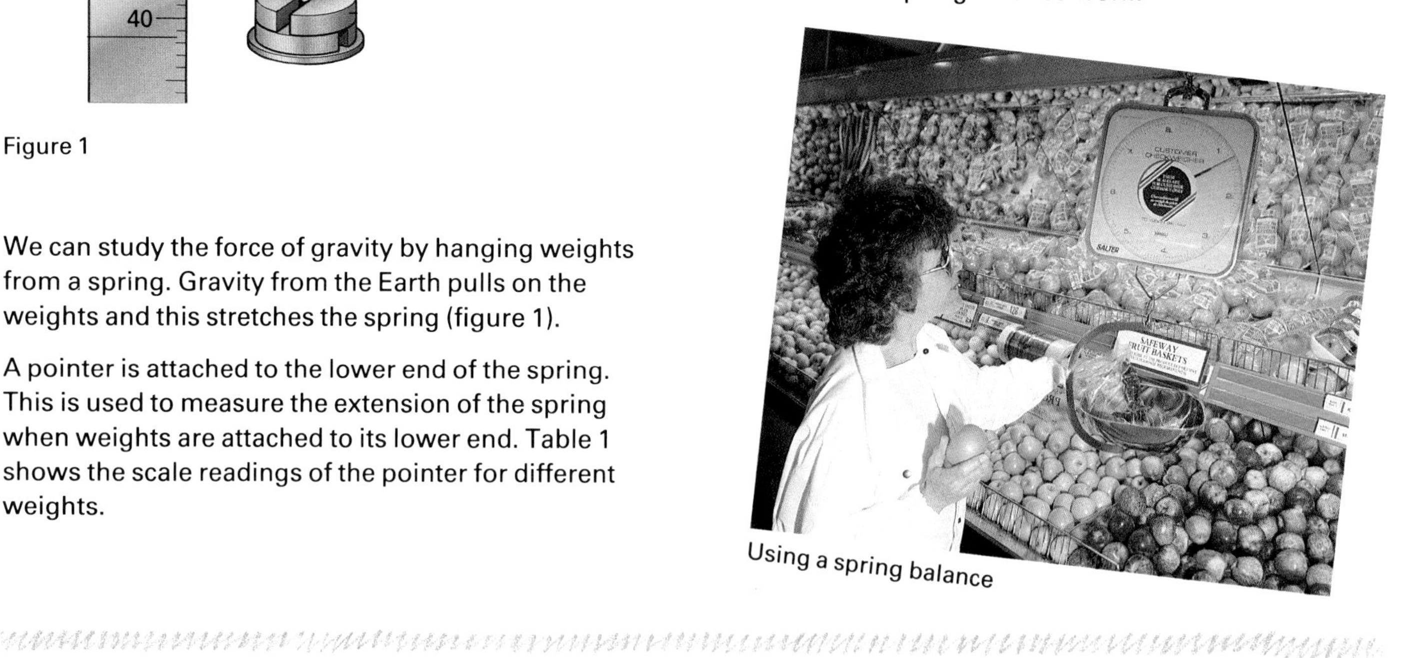

Using a spring balance

5 Water supply

Most buildings in towns and cities get their water from a tower or a reservoir on high ground. The water flows through pipes to homes, offices and other buildings (figure 2).

1 Why is the dam wall thicker at the bottom?
2 Why is the water pressure greater in rooms at the bottom of the tower block?
3 Why is the water pressure sometimes lower in dry, summer months?
4 Why is it necessary to pump water to storage tanks at the top of the block of flats in figure 2?
5 Where does the water supply to your home come from?

6 Make a list of the ways in which large quantities of water are used in your home (e.g. washing dishes).

7 Estimate the volume of water used in your home in one day.

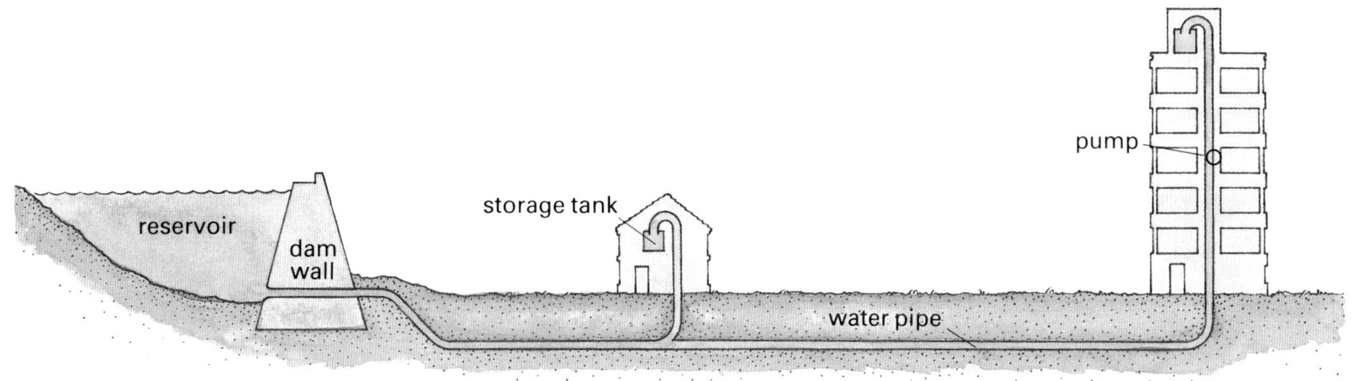

Figure 2 A simplified diagram of a water supply system

6 Designing a parachute

Work in groups of 3 or 4

A toy firm wants to produce a paper parachute for children to use with their toy soldiers.

The parachute must:

(i) fall from a height of 2 metres to the ground in the longest possible time when it supports 3 paper clips. (3 paper clips weigh about the same as one of the toy soldiers.)

(ii) be made from one sheet of A4 paper, or less

(iii) be released at a height of 2 metres during tests. The parachute must not be pushed or thrown.

Imagine that you are designers working for the firm. Your job is to design and make the toy parachute. You are allowed only the following:
3 sheets of A4 paper,
1 metre rule,
10 paper clips,
3 metres of cotton thread,
1 metre of sellotape,
1 pair of scissors,
1 stop watch or clock.

1 Draw a diagram of your model.
2 Give the results of the tests with your model.
3 Compare your parachute and your results with those of other designers.
4 Write a report about the different parachutes and say which gave the best results.

Why is drag important to anyone using a parachute?

7 Turning forces

Forces can change the shape and movement of objects. They can also make things turn. We use turning forces to turn handles. Turning forces also make doors turn on hinges and they help a car to turn corners.

Figure 1

Try the next experiment

1 Push on the edge of a door (point X in figure 1) so that it closes steadily. Open the door again and push on the edge of the door with a greater force.
How does the door move this time?
2 Open the door again. Practise pushing on the edge of the door so that it closes steadily (figure 1).
Open the door again and then push *in the centre* of the door (point Y in figure 1) with the same force.
How does the door move this time?
3 Open the door again. Push on the door with the same force but *near the hinges* (point Z in figure 1).
How does the door close this time?
4 How does the effect of the force depend on where it is applied in relation to the door hinge or **pivot**?

8 Glass bottles or plastic bottles?

Suppose you are the manager of a dairy and you want to modernise your equipment. One decision that you have to make is whether to use plastic bottles or glass bottles. Some of the properties of glass and plastic bottles are compared in table 1.

1 List the advantages of changing to plastic bottles.
2 List the disadvantages of using plastic bottles.

Property	Glass bottle	Plastic bottle
Cost to produce	high	low
Mass of 1 dm^3 bottle full of milk	1.6 kg	1.2 kg
Strength	cracks if knocked breaks if dropped	does not crack if knocked dents if dropped
Effect on taste of milk	none	none
After use	can be returned to dairy for refilling	thrown away, but not biodegradable

Table 1

3 Make a list of any other information that you would like before deciding whether to use plastic bottles instead of glass bottles.
4 Write a report for your board of directors indicating whether you think that glass or plastic bottles are better. Give reasons for your choice.

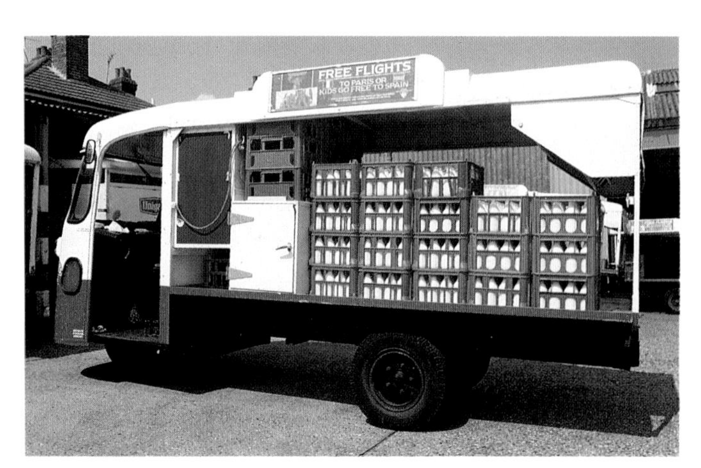

If you had to deliver milk, would you prefer glass bottles or plastic bottles?

9 What makes a bridge strong?

The strength of a bridge depends on two important factors:

- The material used to make the bridge.
- The way in which the material is used.

1 Stone is a strong material for bridges. Why are bridges built from a single stone never very wide?
2 Arch bridges are sometimes made of stone. How are the stones arranged in an arch bridge? Why are the stones arranged in this way?
3 Nowadays, the longest bridges are made of steel and concrete. What are the advantages of these materials?
There are forces acting on most parts of a bridge. Figure 2 shows the forces on a simple beam bridge. The load on the beam is the weight of the girl plus the weight of the beam itself. This load makes the beam bend.

Figure 2

When the beam bends:

- the top surface is being pushed in. The forces are trying to squash it.
  Engineers call this *compression.*
- the bottom surface is being pulled out and stretched.
  Engineers call this *tension.*

Engineers have to design bridges in a way that reduces the forces of compression and tension. The best way to do this is to stop the bridge from bending. Look at the photos above.

4 For each bridge say:
(i) what stops it falling down,
(ii) what stops it bending.
5 Draw a sketch of the suspension bridge. Suppose there is a heavy load at the centre of the suspension bridge. Colour in red those parts of the bridge which are in compression. Colour in blue those parts of the bridge which are in tension.
6 Reinforced beams are often used in bridges. These are made of concrete with steel rods or girders through them (figure 3).
(i) Why does the steel make the bridge stronger?
(ii) Why is the steel often positioned at the bottom of the beam?

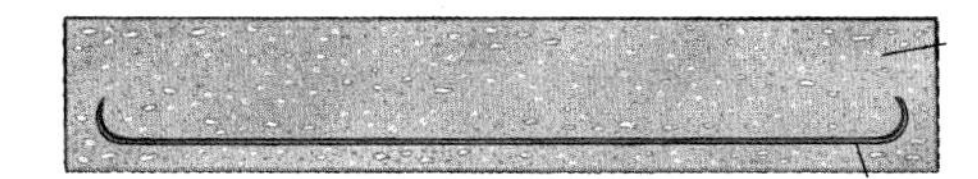

Figure 3

10 What are bridges made of?

Hundreds of years ago, bridges were usually made of wood. Plenty of wood was available and it was cheap. However, the wood slowly rotted away and it could be easily damaged by fire.
From the beginning of the 18th century, stone was used more and more for bridge building. There were good supplies of stone in most areas. It was also strong, cheap and very resistant to fire and corrosion. But the quarrying, cutting and shaping of the stone required more time. From the beginning of the Industrial Revolution in the late eighteenth century, steel became an important material for bridges. Steel was strong and fairly cheap and it could be cast into the required shapes. Its main disadvantage was that it needed to be painted regularly to stop it from rusting.
Nowadays, reinforced concrete is usually used for large bridges. In reinforced concrete the concrete is strengthened with steel rods running through it. Concrete is cheap, it is resistant to fire and corrosion and is fairly strong.

Read carefully through the article above.
One factor which influences the material used for a bridge is cost. What other factors can you find mentioned in the article?

A concrete road bridge over the A21 in Kent

The Tyne and Swing Bridges at Newcastle

OLD LONDON BRIDGE

London Bridge in 1600 A.D.

11 How strong is hair?

Elisabeth has long brown hair and Imdad has short dark curly hair. They decide to carry out a test to see whose hair is strongest. They set up the apparatus below.

1 Describe an experiment that Elisabeth and Imdad might carry out using this apparatus to see whose hair is the strongest.
2 Write down *three* things they should do to make the experiment a fair test.

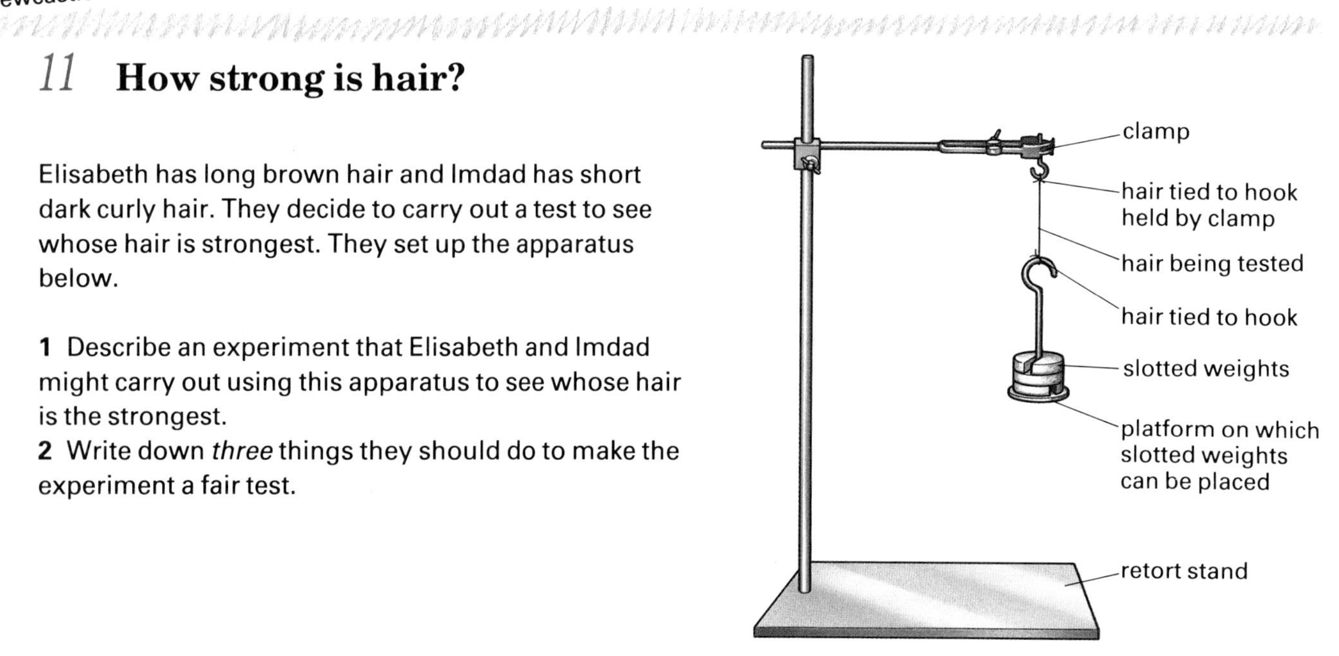

Figure 1

12 Bridging the gap

A log bridge in Burundi, Africa, which shows how a simple bridge can be put to good use

Work in small groups.

Position your desk or table so that one edge is 15 cm from the edge of another desk or table. Now, using only one sheet of A4 paper, build a bridge to span the 15 cm gap (figure 2). The diagram below might give you some help.

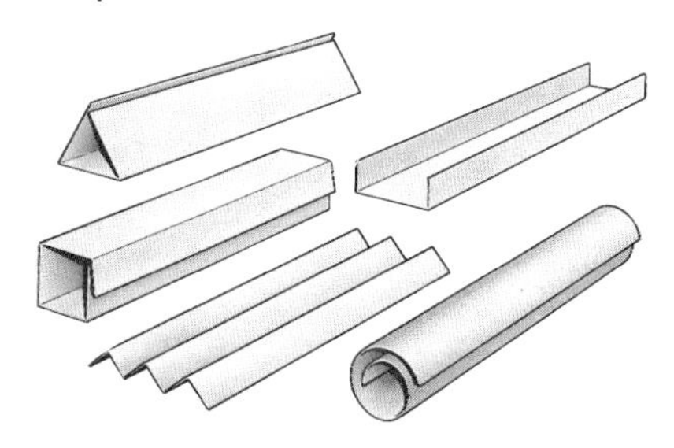

The bridge must support a mass of at least 200 g at its centre. How strong can you make your bridge? You may use glue or small pieces of sellotape to hold the paper bridge together. But, you must not use glue or tape to fix the bridge to the edges of the gap.

1 Explain how your bridge is made. You may wish to draw diagrams.
2 What is the largest mass which your bridge can support at its centre?
3 Use a second sheet of A4 paper to build another bridge. Try to make your second bridge stronger than the first.
4 Compare your bridges with those of others.
5 Write a report about the different bridges. Say which bridge was the strongest and which design features helped to make the bridges stronger.

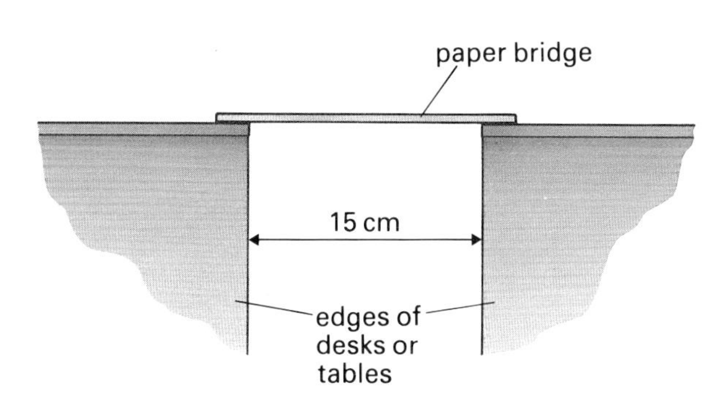

Figure 2

13 Why do steel ships float on water?

1 Why does a block of wood float on water?
2 Why does a steel rod sink in water?

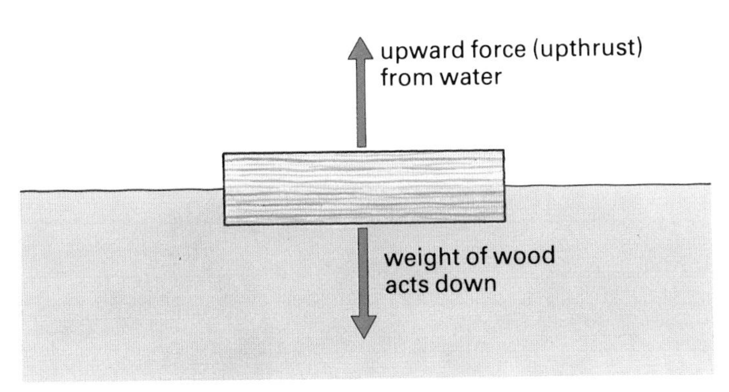

Figure 3 Balanced forces on a floating block of wood

When objects float on water, the weight of the object acting down is balanced by an upward force from the water. The special name, **upthrust**, is used for this upward force (figure 3).

The next experiment will help you to understand the idea of upthrust.

3 (i) Sprinkle some sand into water. Explain what happens. Put a spoonful of sand in a test tube and then set it floating in water.
(ii) Find out how much sand you can put in the test tube before the test tube just sinks.
(iii) Find the weight of this sand.
The weight of this sand plus the test tube balances the upward force (upthrust) on the test tube when it is just floating.
4 Now try to explain why ships made of steel float on water, even though steel rods sink.

14 Examining wind speed

In January 1990, a severe storm hit Britain. There were gusts of wind with speeds well above 160 kilometres per hour (100 miles per hour). The forces from the winds caused terrible damage.

Storm damage from the hurricane in January 1990

Beaufort number	Wind speed (mph)	Description	Effects produced
0	below 1	calm	smoke rises vertically
1	1–3	light air	smoke drifts a little
2	4–7	light breeze	leaves rustle, wind felt on face
3	8–12	gentle breeze	leaves and small twigs move, light flags extended
4	13–18	moderate breeze	dust raised, small branches sway, flags flap
5	19–24	fresh breeze	small trees sway
6	25–31	strong breeze	umbrellas difficult to use, large branches sway
7	32–38	near gale	trees sway, hard to walk against wind
8	39–46	gale	twigs broken off trees, very hard to walk into wind
9	47–54	strong gale	slight damage to houses, slates and tiles blown off
10	55–63	storm	trees uprooted, serious damage to houses
11	64–72	violent storm	widespread damage
12	above 72	hurricane	disaster, terrible damage

Table 1 The Beaufort Wind Scale, used to measure winds

You can estimate the speed of the wind using the Beaufort Wind Scale in table 1. This scale puts winds in order of strength. What number on the Beaufort Scale relates to wind speeds recorded in the hurricane in January 1990?

Wind speeds are measured using an instrument called an **anemometer**.

The diagram shows a simple vane anemometer. When the wind blows, the vane swings back. The wind speed can be taken from the circular scale.

1 Make a list of materials that you would use to make an anemometer like that in the diagram.
Say what each material is used for.
2 Suppose you need only measure wind speeds up to number 5 on the Beaufort Scale. Describe the main steps that you would follow in making the anemometer.
3 How would you mark the circular scale so that you can use it to find wind speeds. (*Hint*: look at the effects produced by different winds in the Beaufort Scale).

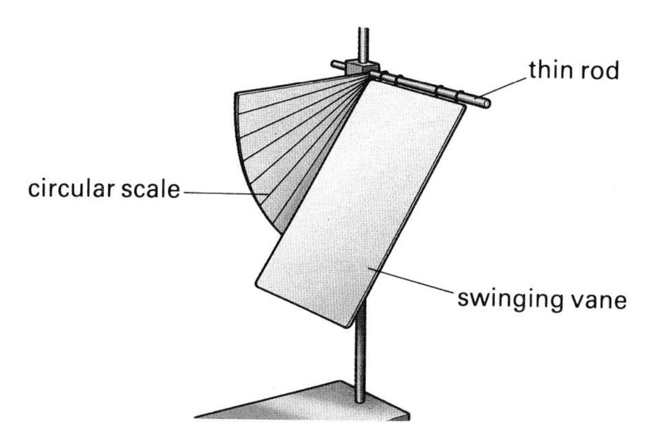

Figure 1 A vane anemometer

4 How would you adjust your anemometer so that it can be used with winds above number 5 on the Beaufort Scale?
5 Make your anemometer.
6 Use your anemometer to measure the wind speeds in different parts of your school grounds.
7 Make a record of your results and try to explain them.

15 Living in space

Get into groups of three or four.

Imagine that you are astronauts preparing for a space flight.

1 Make a list of the problems you will have during a space flight lasting several weeks. Think about the problems that you might have with breathing, sleeping, eating, keeping fit, 'walking' outside the space craft, etc...
2 Discuss ways in which these problems might be overcome.
3 Report your conclusions to the rest of the class.

Astronauts Story Musgrave and Donald Peterson carrying out routine work outside the space shuttle (April 1983)

16 Forces – word quiz

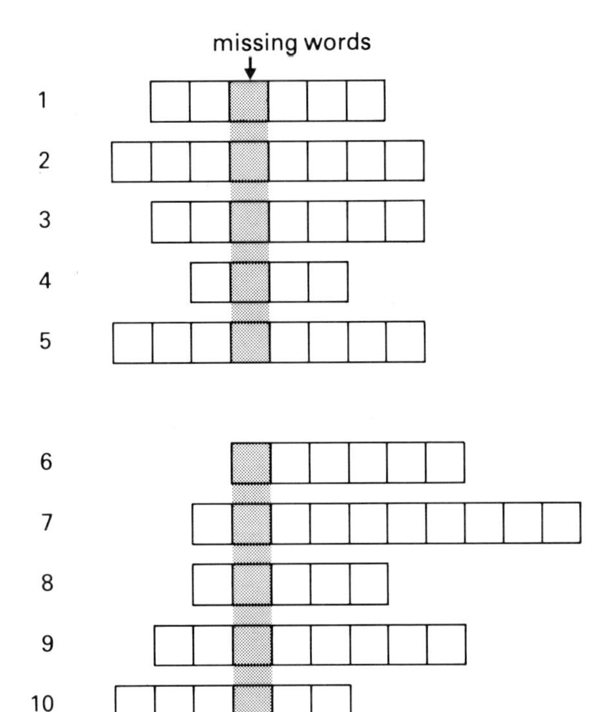

Make a copy of the word quiz below. Answer the clues to find the missing word.

1 The name for your force on the ground.
2 This describes how concentrated a force is.
3 This force pulls you to the ground.
4 The unit for this quantity is kilogram.
5 This hinders movement.
6 The unit used to measure forces.
7 Astronauts are in this state in outer space.
8 This kind of force causes an object to turn.
9 The upward force that a liquid exerts on an object.
10 The unit used to measure time.
11 Two of these forces act in opposite directions and cancel each other.

17 Newton's ideas about motion

About 2300 years ago, the early Greek philosophers studied motion. They thought that:

- objects should not normally be moving
- all moving objects tend to stop moving
- an object will only keep moving if a force is exerted on it.

A stone thrown upwards soon falls back to the Earth and stops moving. A cart rolling on a level road soon stops. The early Greek philosophers thought that the stone and the cart stopped because there were no forces to keep them moving.

In the seventeenth century, Isaac Newton explained these things differently. Newton said:

- objects will not change in motion if the forces on them are balanced
- a moving object will only change its speed or its direction if an extra (unbalanced) force is put on it
- an object at rest will only start to move if an extra (unbalanced) force is put on it.

According to Newton, a stone thrown upwards comes back to Earth because of the force of gravity. A cart rolling on a level road stops moving because of friction. If there were no forces on the ball or the cart, they would keep on moving forever.

1 How did the Greek philosophers explain why a ball rolling on a flat surface stops moving?
2 How did Newton explain why a ball rolling on a flat surface stops moving?
3 How did Newton's ideas about moving objects differ from those of the Greeks?
4 Describe an experiment that you could carry out to show that friction can stop things moving.